S0-BZT-028

HOW CAN I EXPERIMENT WITH ... ?

LIGHT

Cindy Devine Dalton

Cindy Devine Dalton graduated from Ball State University, Indiana, with a Bachelor of Science degree in Health Science. For several years she taught medical science in grades 9-12.

Teresa and Ed Sikora

Teresa Sikora teaches 4th grade math and science. She graduated with a Bachelor of Science in Elementary Education and recently attained National Certification for Middle Childhood Generalist. She is married with two children.
Ed Sikora is an Aerospace Engineer, working on the Space Shuttle Main Engines. He earned a Bachelors of Science degree in Aerospace Engineering from the University of Florida and a Masters Degree in Computer Science from the Florida Institute of Technology.

Rourke
Educational Media

rourkeeducationalmedia.com

©2001 Rourke Educational Media

All rights reserved. No part of this book may be reproduced or utilized in any form or by any means, electronic or mechanical including photocopying, recording, or by any information storage and retrieval system without permission in writing from the publisher.

www.rourkeeducationalmedia.com

PROJECT EDITORS
Teresa and Ed Sikora

PHOTO CREDITS: Page 6: © Galyna Andrushko; Page 9: © OGphoto; Page 15: © digitalskillet; Page 20: © mstay; Page 24: © NASA; Page 27: © letty17; Page 29: © alvarez

ILLUSTRATIONS
Kathleen Carreiro

EDITORIAL SERVICES
Pamela Schroeder

Also Available as:

Library of Congress Cataloging-in-Publication Data

Dalton, Cindy Devine, 1964–
  Light / Cindy Devine Dalton.
    p. cm. — (How can I experiment with?)
  Includes bibliographical references and index
  ISBN 978-1-58952-014-1 (Hard Cover)
  ISBN 978-1-58952-034-9 (Soft Cover)
  1. Light—Juvenile literature. 2. Light—Experiments—Juvenile literature. [1. Light. 2. Light—Experiments. 5. Experiments.] I. Title

QC360 .D35 2001
535/.078 21   00066529

Printed in China, FOFO I - Production Company
  Shenzhen, Guangdong Province

rourkeeducationalmedia.com

customerservice@rourkeeducationalmedia.com • PO Box 643328 Vero Beach, Florida 32964

Light: A natural or artificial bright form of energy that enables us to see.

Quote:

"I find out what the world needs, then I go ahead
and try to invent it."

-Thomas Edison (Inventor of the light bulb)

Table of Contents

Do You Ever Think About Light?

Where does light come from? Where does it get its color? Why is it brighter in some places than others?

You will feel like a genius after you find out the answers to these very good questions. You are about to discover just what light is! Hang on and have fun!

All of Earth's natural light comes from our sun.

Light Is All Around Us

Without light, we couldn't see. When you look at things, what you are actually seeing is light bouncing off the objects. For example, when you look at this page you are seeing the light bouncing off of it.

Sunlight provides energy that helps plants grow. Animals eat plants and we eat animals. Without light, Earth would not have any plants, or animals, or people!

Sunlight helps plants and animals grow. Without light, we wouldn't be here.

Particles of Light

Light is a flow of tiny particles. These particles are called **photons**. A photon is one particle of light. Photons contain the energy that makes up the energy of light. It's easy to picture light as a particle. Imagine yourself throwing a tennis ball at a wall and watching it bounce off. The tennis ball acts like the light particle bouncing off a mirror.

The image to the right shows the distance between, or wavelength, of two different kinds of light waves, one from each end of the spectrum.

Catch the Wave of Light

Have you ever dropped a rock in the pond and watched the waves spread outward? The distance between the high points of the wave is called the **wavelength**. The number of waves that pass in a time period is called the **frequency**. Light is a special type of energy that travels like a wave, called an electromagnetic wave. Different colors of light have different frequencies.

**Wavelength
Red**

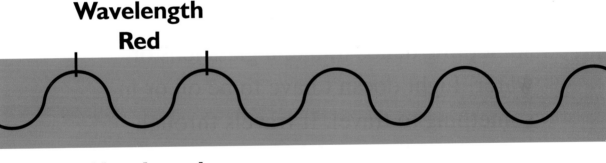

**Wavelength
Violet**

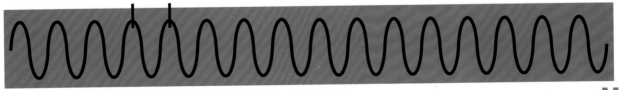

How Fast Does Light Travel?

Light travels faster than anything else. The speed of light is about 186,000 miles per second (300,000 km/s). Light is a type of energy that travels like a wave—a straight wave. Light cannot bend unless something bends it, like a glass lens or water. Light doesn't have to be on or in something to travel. It travels through space and all around us.

The light from this flashlight cannot bend to shine around corners by itself. The glass of water shows a great example of how water can bend light. Look at the spoon.

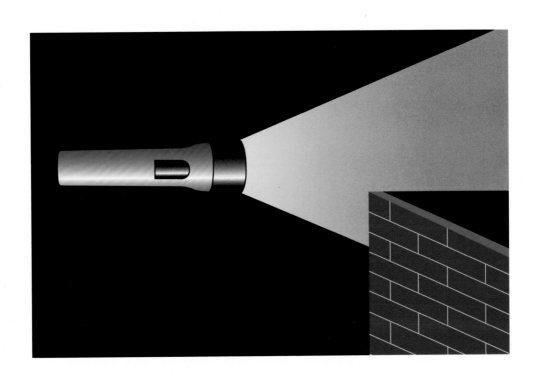

Is the Light Bulb Hot or Cold?

Light comes in several different forms. Look in your school. Do you see round light bulbs or long tube-like light bulbs? When you're camping, what kind of light do you see? Do you see a campfire, a flashlight, or the moon? These are all different types of light that have different energy levels. What energy level is the sunlight? Sunlight has **ultraviolet** light. Ultraviolet light can harm your skin and your health. Ultraviolet light is very powerful because it has a high frequency that is able to pass into skin.

One kind of light gives off heat and one doesn't. Hot light is **incandescent** light.

Ultraviolet light can be dangerous to our eyes and skin. Sunglasses help protect our eyes from harmful rays.

Candles, the sun, and some light bulbs give off incandescent light.

A cool light is **luminescent**. An example is a **fluorescent** light. The long tubes of light in most classrooms are fluorescent. Fluorescent lights do not use as much energy as incandescent lights.

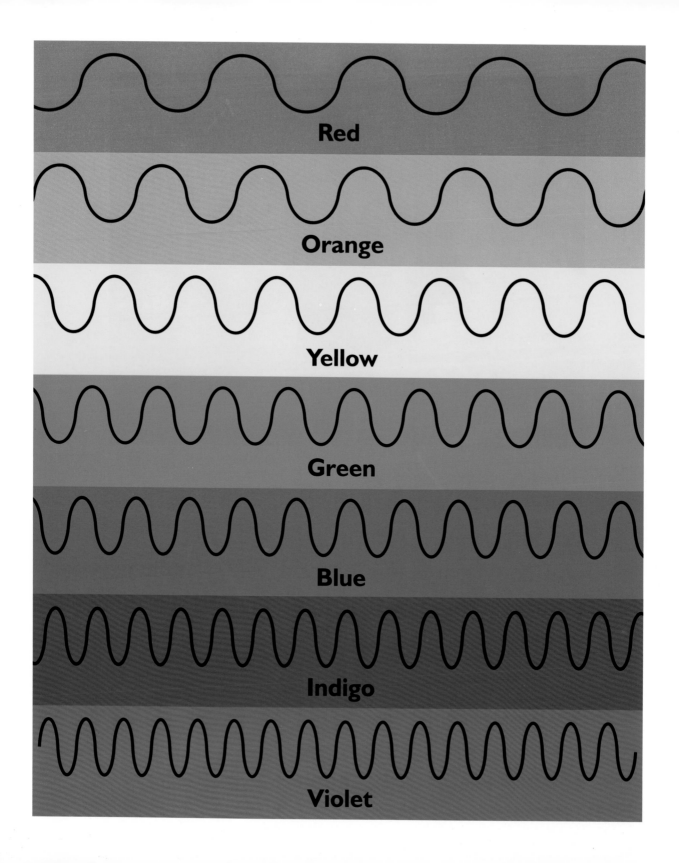

What Color Is Light?

Yellow or white light seem like they are the brightest colors. They're not. The white and yellow that we see in sunlight are actually a mixture of all the colors of light. There are seven colors in the color **spectrum**—red, orange, yellow, green, blue, indigo, and violet—just like a rainbow. Our eyes are most comfortable looking at green. At the Earth's surface, green is the brightest color.

Did you know that a light bulb actually gives off more red and orange light? A fluorescent bulb produces more blue and yellow light.

Each color in the spectrum is measured by its wavelength.

Hands on:

Getting a Sneak Peek at What Light Is Really Made of

What you need:

- A mirror
- A piece of paper
- A bowl or glass of water

Try This:

1. Place the mirror in the glass or water at an angle. (Let it lean against the side of the glass.)
2. Turn the glass or bowl so the mirror is facing the sun.
3. Hold the paper at a slant in front of the glass. Move the paper around until you see the rainbow of colors. You may need to move the paper around slightly until the colors come into full focus.

Here you can see the rainbow of colors reflected onto the paper.

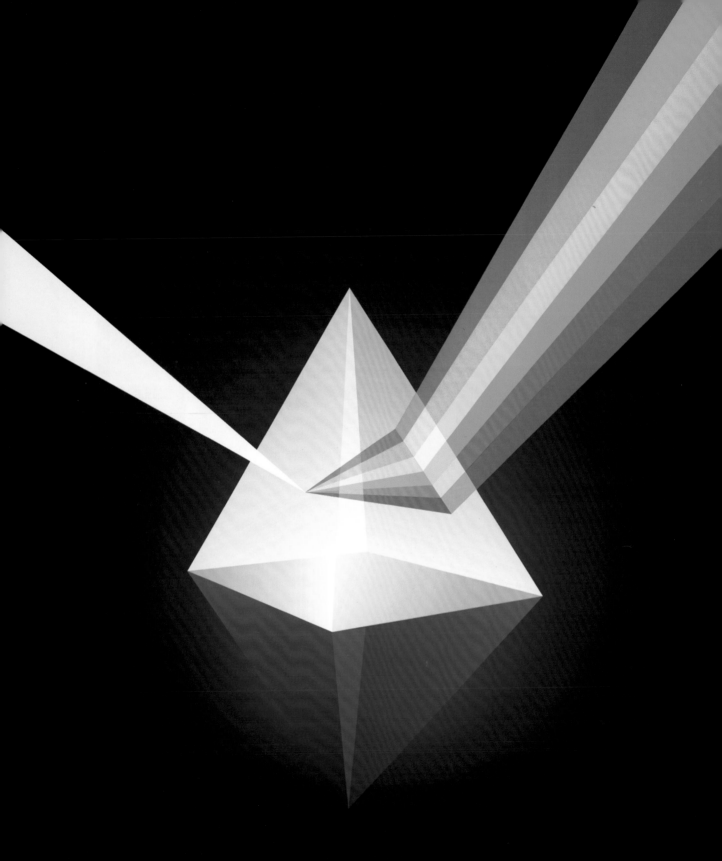

What Happened?

Light looks white, but it actually isn't. What you are seeing is the waves of light together. When light passes through the water and bounces off the mirror, the light waves are broken apart and you see seven different colors. The same thing happens when light passes through a **prism**.

When light passes through a prism the light waves are broken apart, showing seven different colors.

Hands on:

Sky In A Jar

What you need:

- A clear, straight-sided drinking glass or clear plastic or glass jar
- Water
- Milk
- Measuring spoons
- Flashlight
- Darkened room

1. Fill the glass or jar about 2/3 full with 8-12 oz. (250-400 ml) of water.
2. Add 1 teaspoon (2-5 ml) of milk and stir.
3. Shine the flashlight above the surface of the water and observe the water in the glass from the side. It should have a slight bluish tint.

4. Hold the flashlight to the side of the glass. Look through the water directly at the light. The water should have a reddish tint.

5. Put the flashlight under the glass and look down into the water from the top. It should have a deeper reddish tint.

What Happened?

The small particles of milk in the water scattered the light from the flashlight, like the dust particles and molecules in the air scatter sunlight. When the light shines in the top of the glass, the water looks blue because you see blue light scattered to the side. When you look through the water directly at the light, it appears red because some of the blue light was removed by scattering.

Crazy Questions About Light

Question:

If I could fly in a spacecraft at the speed of light, what would happen?

Answer:

Time would slow down for you. You could reach the center of the Milky Way in your lifetime! However, if you decided to come home, thousand of years will have passed!

The Milky Way is made up of more than 100 billion stars.

Question:

How does sunlight make leaves green in the summer, red and orange in the fall and brown in the winter?

Answer:

Sunlight provides energy for plants through **chlorophyll**. When plants receive a lot of sunlight the chlorophyll in the plants turns green. When they receive less, leaves turn red and orange. When they do not receive enough to stay alive, they turn brown and often fall to the ground. So when days are long, as they are in the summer, leaves are as green as they can be. The less daylight there is, the less bright the leaves will be.

In autumn months there is less sunlight than in the summer months. That is why leaves turn color in the fall.

Question:

Why do rainbows only come out when it rains?

Answer:

Rainbows don't "come out" in the rain. The raindrops in the air act as tiny prisms. Light enters the drop, reflects off the side of the drop, and exits, showing a spectrum of light.

Question:

What is light pollution?

Answer:

Light pollution is when the sky is lit up by the city lights so much that the night sky cannot be seen. Light pollution makes observing the stars difficult.

The water droplets from this waterfall act like tiny prisms, showing a spectrum of light and creating a rainbow.

Glossary

chlorophyll (KLOR eh fil) — green chemical found in plants that needs sunlight to make energy for plants

fluorescent (floo RES ent) — light that is created by passing electricity through a coating on a tube-shaped bulb

frequency (FREE kwen see) — how quickly something happens over and over again

incandescent (in ken DES ent) — white or glowing with great heat

luminescent (loo mi NES ent) — a form of light that does not give off heat

photons (FOH tonz) — energized particles of light

prism (PRIZ em) — a tool that breaks white light into a full spectrum

spectrum (SPEK trum) — all the colors of light: red, orange, yellow, green, blue, indigo, and violet, arranged in the order of their wavelengths

ultraviolet (ul tra VY eh lit) — violet end of the spectrum; very short wavelength of light

wavelength (WAYV length) —distance between two points in a cycle of waves

Websites to Visit

www.howstuffworks.com
www.sciencemadesimple.com
www.exploratorium.com

Index